British Oil Policy: A Radical Alternative

A Report based on a study made
for the Department of Energy
and submitted to the Secretary of State in March 1979

by

Peter R Odell
Director, Economic Geography Institute,
Erasmus University, Rotterdam, The Netherlands

Kogan Page

Copyright © Peter R Odell
First published in Great Britain by
Kogan Page Limited 120 Pentonville Road
London N 1
ISBN 0 85038 316 1 ✓

Based upon 'The Exploitation of Britain's Offshore Oil Resources'
© Crown Copyright with the permission of the Controller of
Her Majesty's Stationary Office

Printed in Great Britain by
McCorquodale (Newton) Ltd.
Newton-Le-Willows, Lancashire

This book is to be returned on
or before the date stamped below

POLYTECHNIC SOUTH WEST

ACADEMIC SERVICES
Plymouth Library
Tel: (0752) 232323
This book is subject to recall if required by another reader
Books may be renewed by phone
CHARGES WILL BE MADE FOR OVERDUE BOOKS

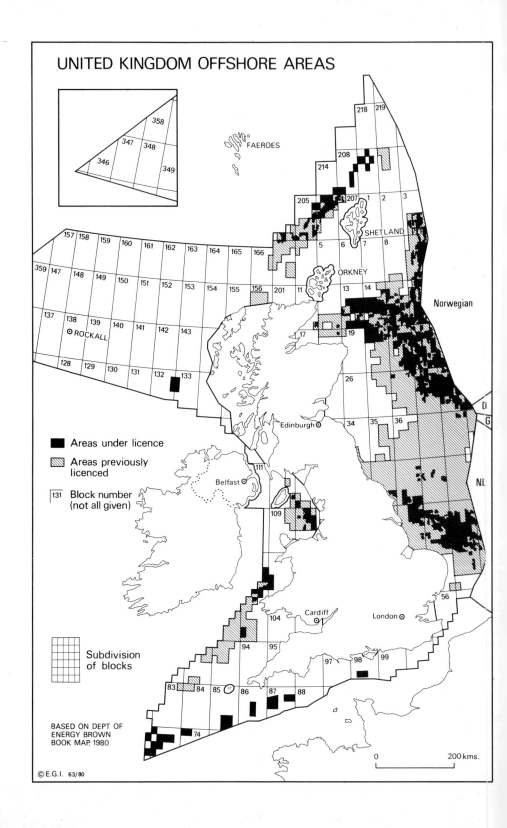

UNITED KINGDOM OFFSHORE AREAS

358
347 348
346
349

FAEROES

218 219

208

214

205 207 1 2 3

SHETLAND

5 6 7 8

157 158 159 160 161 162 163 164 165 166

ORKNEY

359 147 148 149 150 151 152 153 154 155 156 201 11 13 14

Norwegian

137 138 139 140 141 142 143

⊙ ROCKALL

17 19

128 129 130 131 132 133

26

D
G

Edinburgh ⊙

34 35 36

NL

111

Belfast ⊙

109

■ Areas under licence

▨ Areas previously licenced

131 Block number (not all given)

104

Cardiff ⊙

London ⊙

56

Subdivision of blocks

97 98 99

83 84 85 86 87 88

94 95

74

BASED ON DEPT OF ENERGY BROWN BOOK MAP, 1980

0   200 kms.

© E.G.I. 63/80

Contents

Preface

This short study was undertaken as one of two responsibilities given to me during my temporary period of appointment as a consultant to the Department of Energy in 1977/78. The other task was an analysis of the so-called Annex B's to the oilfield development plans submitted to the Department of Energy on a strictly confidential basis by the companies concerned. In order to be able to undertake that analysis, I had to sign a set of confidentiality undertakings — one for each of the companies involved as a partner in one or more of the 13 fields which were under development at that time. In consequence no part of that analysis can be included in this study which is based on published material and on unclassified documentation in the Department of Energy. No confidential material was made available to me from other Departments of State concerned with oil, from state-owned enterprises such as the British National Oil Corporation, or from individual oil companies and consortia of companies working on specific North Sea projects.

The original version of the report was presented to the Minister in March 1979. The government of the day was defeated in the House of Commons at about the same time and it was thus not until several months later that the new Minister was able to take his decision on the report. His decision was to place a copy of it in the library of the Houses of Parliament so that it became a shadowy public document (with Crown copyright) in July 1979. It has since been of modest use to a small number of Members of Parliament (see, for example, the House of Commons debate on the Petroleum Revenue Tax Bill, *Hansard*, 10 December 1979, Cols. 990-1035) and of no use to anyone else.

However, as the report does contain ideas which challenge the

7

validity of several aspects of oil policy and in that it offers suggestions for a radical change in attitudes to, and the organisation of, the exploitation of the UK's offshore oil resources, this revised version has been prepared for publication by Kogan Page Ltd (with permission of HMSO). Over the last year — since the original version was completed — no development in the international world of oil nor in the UK oil policy has, in my view, undermined the essential validity of the report's arguments and conclusions. On the contrary, the politico-economic changes since then make a fundamental reappraisal of the motivations and methods for the exploitation of Britain's offshore oil and gas even more appropriate — and necessary.

The processes of offshore exploration and exploitation are still less than two decades old and most of the offshore potential remains unexamined. The processes will continue for at least another half-century. There is therefore every justification for an evaluation as to the appropriateness of the policies that have emerged from the initial period of exploration and development and for changes to be introduced. This is vital in order to ensure that developments proceed to their full extent and that they bring the maximum possible benefits to the country.

Finally, a word of thanks to all who helped me in the preparation of this report — particularly in the Department of Energy. They would, I am sure, prefer to remain anonymous as the packet of ideas I have put together does not coincide with the packets they would individually prefer. My thanks also to Mrs. E. van Reijn-Herscheit for typing the revised version and to Philip Kogan and Ethel de Keyser of Kogan Page Ltd. for their support and help in this modest publishing venture.

Peter R. Odell
Rotterdam, February 1980

A. Introduction

1. The aims of this report are modest. First, it aims to highlight what appear to be the most important issues involved in the continued development of UK offshore oil. Second, there are recommendations for action in respect of strategic and tactical policy questions.

2. The development of the supply of offshore oil (and of gas), the cost at which it (they) can be produced and the price at which it (they) can be marketed, clearly constitute the main elements determining the shape of the energy sector of the UK economy in the 1980s and the 1990s. Oil and gas in general, and British produced oil and gas in particular, are the energy sources preferred by most British energy consumers. Alternative sources of energy such as coal and nuclear power are really only relevant in the context of the need and ability to supplement oil and gas. This is a clearly defined consumer preference. I would suggest that it should be thwarted only by energy policies, applied with great caution, which seek to limit the contribution of British oil and gas to the energy economy. This is of immediate importance in light of the very much higher resource costs which are involved directly in developing the supply of alternative energies. It is even more important if it is also borne in mind that there would be considerable additional costs connected with infrastructural and other changes in society, in order to enable it to absorb relatively more energy in forms other than oil and gas, on which, by 1979, all main users in the socio-economic system (with the single exception of the electricity generating industry) have become largely dependent.[1]

1. This comment is not intended to imply that UK energy sources other than oil and gas should not be developed or encouraged.

3. Such a view of the way in which the British economy could and should evolve does simplify, to some degree, the study of energy policy issues (most notably because it defines where the main effort should be directed, at least for the immediate future). However, it does not, by any means, eliminate all the difficulties facing the country's energy planners. Quite the contrary: in analysing and defining the role of oil and gas there is a set of geological, technological, economic and political influences at work which serve to create problems of comprehension, let alone of evaluation. This report seeks to expose these problems in a systematic way. The difficulties in attempting to integrate oil policy are apparent and I shall not be unduly surprised if there are 'holes' in the presentation or, even worse, internal inconsistencies in the arguments. I hope, however, that these will not be sufficient to undermine the validity of the report. Indeed such criticisms could produce a basis for the improvement of the analysis and the recommendations.

The output and development of the coal industry, in particular, ought, in my view, to be maintained. This would necessitate a continuing programme of new production ventures and of research and development on the future use of coal. However, I see this, at the moment, essentially as an insurance policy for the post-1995 period in case resource-based limitations mean that the country's oil and gas-based economy cannot be sustained thereafter.

B. External politico-economic influences: the necessary responses

1. Current international economic and political influences on oil developments are particularly difficult to analyse and evaluate, in a world which is still going through a revolution in the mechanisms for determining the supply and price of oil. It seems, therefore, that the opportunities for most effectively developing the oil resources of the UK Continental Shelf have, to date, been judged only in the context of what has become a conventional, rather simplistic view of the future of oil, *viz* that the commodity will inevitably become scarce and increasingly expensive over the time horizon at which it is necessary to look for planning purposes. For example, in the Department of Energy's *Energy Policy Green Paper (Energy Policy: A consultative document*, HMSO, 1979) we read:

> 'Even the most optimistic forecasts see oil supplies levelling off and then falling before the turn of the century and the most pessimistic sees demand overtaking supply as early as 1983. Perhaps the most likely outcome is that world oil supplies will begin to level off in the late 1980s, reach a peak in the early 1990s and decline thereafter' (pp 11-12)

2. The Green Paper further assumes 'for the purposes of the forecasts' that oil prices will rise gradually to around $20-25 per barrel (in 1977 dollar terms). It should be noted, however, that this does not involve an annual rate of increase in the real price of oil of more than three per cent. Under almost any circumstances this is a lower percentage figure than one would have to use to discount the value of oil reserves which are deliberately kept unproduced for use in the distant future. Thus even this assumption does not, in itself, provide an effective argument for a restrictive approach to the depletion of the UK's

offshore oil resources. Nevertheless, this sort of view on the future supply and price of oil is used to justify a conservationist approach to oil resources. Don't produce too much too quickly, it is argued. It is better to save oil for an uncertain future — or rather for a future in which oil will inevitably be in short supply.[1]

3. As a consequence of this kind of argument pressures have built up for the introduction of restrictions on the rate of output of individual fields. In addition, it is considered that plans for the rate of exploration and development of the rest of the UK Continental Shelf should be cut back. The impact of such unjustified pressures constitutes a handicap to the effective exploitation of offshore oil. First, there is the adverse effect the results of such pressures would have on the overall attractiveness to the oil companies of expenditure in the UK's oil provinces, relative to their opportunities in many other geological provinces around the world. Second, there are the effects that such pressures have on the economics of producing an individual field. They not only have an unfavourable influence on the relationship between the discounted value of the oil and the levels of investment needed to produce it but they also increase the investment risks. This is because, in the longer term, the companies will not be able to use the production and transport installations, which they have built in the harsh environmental conditions of the North Sea, for near-future production. If production is delayed then there is a risk that investment in the infrastructure facilities will be wasted.

4. The adverse consequences of a restrictive approach to the exploitation of Britain's offshore oil need not, moreover, be limited to the exploiting oil companies. They may also affect the national interest. This is because the approach can lead to the failure to maximise the present value of revenues from oil production. It can also lead to a limitation on the favourable balance of trade arising from offshore developments. Neither the documents to which I have had access nor the discussions

1. If oil is going to be in 'short supply', however, then the price over the next 20 years would increase to very much more than the maximum of $25 per barrel specified in the planning assumption. There appears, in other words, to be a serious inconsistency in the *Energy Policy* presentation.

that I have had within the Department of Energy have convinced me that effective evaluations to test for the impact of this important consideration have yet been made.

5. This represents only one element of doubt about the validity of the present policy strategy for the development of the country's offshore oil resources. And it may, indeed, be the minor element. An even more important aspect is the relatively high probability (over 50 per cent in my view) that the official interpretation of future developments in the international oil supply and price situation (*viz* steadily increasing real prices as a result of an inherent scarcity of the commodity) is incorrect.[1] Given this belief in scarcity, the prospects for UK oil development can be assumed to be independent of what is happening elsewhere in the world because developments in the international situation will never impinge on the ability of all British oil to find a market at continuously increasing real prices. This conclusion seems to be accepted without question as a basis for policy making. But such an approach is not necessarily justified, given the following considerations:

(a) UK oil is, to a very large degree now, and to a large degree

1. It would make this report too long and remove it from its main theme if I were to advance my alternative views in detail here. I have recently done this in the final chapter of the revised 5th Edition of *Oil and World Power* (Penguin Books, Harmondsworth, 1979). The essence of my argument is that the world's potential oil (and gas) resources are at least twice and possibly up to four times larger than the quantities which are assumed in the 'inherent scarcity' hypothesis. Moreover, as the demand for oil is now growing much more slowly (at less than 1 per cent per annum instead of at 7-8 per cent) even the less optimistic view of future oil resources would enable the industry to continue to expand well into the second quarter of the 21st century. Current constraints on oil resource developments are thus nothing to do with the ultimate size of the resource base. Rather they are technico-economic (in respect of the difficulties associated with working in new habitats for oil) and institutional (in respect of the unwillingness and/or inability of the oil companies to operate in most parts of the world where potential resources thus remain undeveloped). However, both sets of constraints are capable of being removed, thereby opening up the prospects for plentiful supplies of oil over the whole period during which the world's economic systems are likely to want to use oil in increasing quantities: a period which one may assume will be terminated sometime in the first half of the 21st century when alternative energies become available on a large scale and make oil a less economic source of energy.

13

for the foreseeable future (in the absence of important policy changes), fully integrated into the supply/refining and distribution networks of the major international oil companies. These may or may not want, or indeed be able, to absorb all the UK oil they could produce. This is dependent upon the results of alternative supply, refining and marketing options open to them and on the basis of which they will seek to optimise the returns from their operations at the international level, to the detriment of some national interests.

(b) Sometime before 1985 the major oil exporting countries seem likely to have to take quota or even pricing action to ensure that their exports secure preference in markets which will become weak, because of the actual or the potential over-supply of oil in a severely constrained demand situation. The Western European oil market seems the one which is most likely to be approached in this way by member countries of OPEC. This is because the demand for oil in Western Europe now seems unlikely to grow very much; because it is the area in which competition from British and other non-OPEC oil is likely to be felt most strongly; and because many OPEC countries are building up a high degree of inter-dependence with Western European countries (in trade, commerce, finance, investments etc). This inter-dependence is being developed so that the OPEC countries will be able to use those relationships to ensure that they can achieve their oil export plans. European countries will thus become susceptible to pressures to take specified quantities of oil from the OPEC countries on which significant sectors of European industry and commerce, to a large degree, now depend.

(c) The UK appears, so far, to have made no efforts to ensure guaranteed long-term outlets for its own rapidly increasing oil production. The oil companies do not seem to be required to do this and BNOC appears not to want to do it.

(d) Although it is not the most likely development, there is, nevertheless, a chance of an international oil market collapse in the later 1980s — the period with which UK oil policy is now most concerned. This arises from the prospects for important new sources of supply (including the

North Sea, Mexico, China and North America) in a situation of extremely slow growth in the demand for oil — partly as a consequence of deliberate conservation measures and partly from low rates of economic development. As a result an OPEC member may become unwilling or be unable to stick with the organisation's minimum oil price decisions. If this happens it will place at risk a part of Britain's existing capacity to produce oil. Some UK oil would not be worth producing compared with alternative supplies available; it would reduce the value of the oil that is produced with both balance of payments and government revenue implications; and it would undermine most of the potential for continuing exploration and development activities on the so far undeveloped and/or unexplored parts of the UK Continental Shelf. Such exploration and development would be seen as too high cost and/or too high risk potential to make the necessary investment worthwhile.

6. Thus there seems to be the need for a study of the relationships of various possibilities for the future international oil situation to the prospects for the continued, successful exploitation of the oil resources of the UK Continental Shelf. Such a study would need to deal with the many variables involved with the object of:

(a) Evaluating what alternative options are open to the UK in its offshore oil development in the light of contrasting international developments in the world oil market;

(b) Indicating ways of protecting British oil wealth *against* a possible deterioration in the world oil price in general and/or an attempt by other oil-producing countries to 'corner' part of the markets which are important for UK oil;

(c) Suggesting mechanisims which will ensure that future decisions on how much British oil to produce are not made by the producing companies solely by reference to their concern for optimising the overall returns on their operations at the international level. How, and how far, in other words, can decisions on British oil be divorced from internationally orientated corporate decisions?

(d) Indicating how the *use* of UK oil production (ie its use in the world's markets) can optimise British, rather than individual company interests and/or the interests of the countries in which the companies have their headquarters.

There is, therefore, a requirement for positive UK strategic oil planning, so that the country is not simply obliged to react, or even to bow, to external pressures. Instead, UK oil policy should aim to be influential in helping to determine the global oil situation for the next critical 10 to 15 years. Current energy policy is based on the belief that there is little the UK can do to affect the world price of oil, or the development of the international oil situation. This is not only defeatist: it is also untrue. The UK, through a policy which ensures its effective control over some of the most significant of the world's oil reserves, in terms of their location in the centre of the world's most energy-intensive using region and in terms of their occurrence in a region which is politically stable, could evolve a strategy which would be of crucial importance in helping to stabilise the international oil situation. Moreover, in using its oil in this way the UK could also help to ensure its own wellbeing in an uncertain world.

C. The national politico-economic environment and the exploitation of offshore oil resources

1. The results of the traditional concession system

(a) Over the last 10 years new and expanded legislation and an increasing number of regulations have served to 'tighten up' the freedoms originally enjoyed by the concessionary companies in respect of exploration, field development and, of course, the tax obligations of the successful companies. I do not propose to define what was 'wrong' with the system to start with or even to analyse the effects that individual changes in the concession and the tax arrangements have had on the system over the years. Such exercises have been undertaken and there is a high degree of awareness of the results.

(b) An examination of the evolution of the system and of its present state persuades me, however, that the government still lacks the essential means to initiate and to guide developments once concessions for exploration have been granted to individual companies or consortia. In essence, we appear to have arrived at the point where the government can — and, indeed, sometimes does — stop things happening but it is still largely unable to cause things to happen. We thus have an essentially negative government approach to the development of oil on the continental shelf.[1] This is the case in respect of several different aspects of the concession system as it has evolved to date. These include:

1. For the moment I am discounting the significance of the British National Oil Corporation (and the British Gas Corporation) in this respect. Their significance — and the limitations on their importance — are dealt with later in the report.

(i) The work programme and the speed of its implementation on the concessions granted

In theory the government, on granting a licence, requires a specified work programme to be undertaken. In practice, this cannot be enforced as a company unwilling to complete a programme may simply hand back a block. Thus, 38 'obligatory' wells from the Fourth Round of Licences have not been drilled.

A failure to drill a well sometimes appears to be accepted by the government because 'new geological evidence' shows that further exploration would not, according to the company, be worthwhile. But such decisions run counter to the requirements of the licensing procedures. A company secured a particular block partly because it promised a specific work programme. It thus ought not to be permitted to withdraw without, at least, paying a penalty equal to the investment it has not made; nor should it be allowed to retain an interest in any part of a licence for which the originally required work programme has not been fully completed. The British North Sea and the rest of the country's continental shelf is, like the rest of the oil-producing world, full of surprises, and a well not drilled is a penalty to the nation. It has been deprived of information from the drilling and of the expenditure on the project with an adverse effect on jobs and income — both directly and indirectly.

Moreover, if 'disappointing new geological information' enables a company to withdraw from an obligation, then encouraging additional geological information in respect of another block, should give the government the right to stipulate immediately wells additional to those to which a company committed itself under its work programme obligations. This should be irrespective of the company's attitude, which may not necessarily be positive, in the light of commitments and opportunities in other blocks in which it has an interest. If the company declines, then the concession for the block should be withdrawn on payment of compensation to the company equal only to the cost of the work it has done on the block.

(ii) *The question of the initiation of developments in respect of successful discovery wells*

The government, in essence, still appears simply to sit and wait for a company to take the initiative on field development. The company's decision depends not only on the inherent value of the discovery itself, but also on what else the company has under way or under consideration in the UK Continental Shelf and what else it has under way in the rest of its international corporate world. But concession agreements made under the 1975 Petroleum Act include the right of the government to order a development, if it judges this to be desirable from the national point of view. A failure by the company to agree to develop, following such a government decision, means that the company's interest in all or part of the licence can be revoked.

A large number of discoveries are not being developed, possibly because of the uncertain economic viability of new field developments from the companies' points of view, or because the companies concerned prefer to concentrate their activities elsewhere — either within, or worse still, outside the British sector. In view of this one wonders why the government has not yet chosen to exercise its rights in respect of ordering development. This would be a more appropriate means of eliminating the risk of a possible decline in production after the mid-1980s than the means proposed by the government — that of depletion controls on fields already in production.

(iii) *The nature of the development programme for a field*

Government involvement here to date seems to have been mainly restricted to a go/no-go response: to questions of timing, and to issues of an infrastructural nature (eg decisions on how best to land the oil from a particular field). This is plainly negative rather than positive government.

It is a problem which will, though only in part, be resolved by much increased state participation (BGC and BNOC). However, even in respect of largely autonomous state corporations, 'prodding' by government may be necessary to secure a field development which is considered desirable from the national and/or the regional point of view. This may be seen, for example, in respect of the long delayed development of the Morecambe Bay gasfield. In this case the BGC's marketing policy has been in conflict with the economic needs of the depressed Liverpool and north-west England regions. Their economies would have benefited, both directly and through local multiplier effects, by a decision to develop the field earlier and, even more, by a decision to use the newly available gas to stimulate invest-ment interest in the region — as in the case of Groningen in the Netherlands.

There are several specific reasons why the Department of Energy should be positively involved in development decisions. These include the need to ensure that the whole rather than a limited part of a reservoir is developed; and/or that all adjacent and overlapping, but geologically separate, pools are similarly incorporated into a com-prehensive development plan; and to make sure that the development plan chosen (it can be one of many considered for a particular field) is the one likely to make the maxi-mum possible contribution to government revenues and/or foreign exchange earnings (when measured in present value terms). The greater the positive government involvement in the long-term development decision for a field, the greater the government's influence in ensuring that the plan is achieved. It also diminishes the degree of company uncertainty over its future prospects from its exploitation of the field. This seems likely to produce a better overall environment for effective government/company co-operation.

(iv) The possible integration of field transport systems with one another

Steps towards achieving such integration between transport systems in appropriate locational relationships appear not to have been taken by the Department of Energy to date. Such action is possible under the 1975 Petroleum Act and there are examples in the North Sea where the failure to integrate the separate interests and operations of two or

20

more groups has led to decisions which appear to be sub-optimal from the national point of view. This arises because field developments have been delayed and/or limited either because 'space' could not be obtained in a separately developed pipeline at a price which made field development possible, or because companies, with contrasting interests in exploiting near adjacent newly discovered reserves, were unable to agree on the joint transport system developments which were a prerequisite for the achievement of the necessary economics of scale.

In order to avoid these situations the Department of Energy needs to exercise its legal right to investigate such situations as a matter of course and, where appropriate, to order joint enterprises. This would make a positive contribution to the evolution of the UK's oil production and transport systems.

(v) *The disposal of British oil to third parties (as purchasers of the crude), to refineries and to other countries*

Existing legislation requires British oil to be landed in the UK (except by special permission of the Secretary of State). In addition, there is an 'expectation' by the government (which is not enforceable in a legal sense under present legislation) that up to two-thirds of UK oil be refined in Britain. The 'control' exercised by this expectation is essentially voluntary and so provides a very modest, if indeed any, constraint on companies' decisions on where to refine their British sector North Sea oil production. Moreover, as far as the former control (ie the UK landing requirement) is concerned, it neither controls *where* it is landed (except by means of physical planning controls) nor what happens to it once it has been landed. In other words, there is no rationally evolved and comprehensive government-regulated integrated 'downstream' system for the disposal of oil from the UK Continental Shelf to ensure that the national interest is taken into account. Simply leaving BNOC to take care of this is inadequate. It handles only part of the oil involved and it needs the Secretary of State's permission to undertake any oil refining. It is, therefore, limited to trading crude oil rather than having the greater flexibility of an integrated oil company.

(c) Even after taking account of the UK Continental Shelf holdings of BNOC and the BGC and of the impact of the negotiated participation agreements, the concession system as developed to date means that the ownership of a large part of the oil being produced — and to be produced in the future — lies, and will continue to lie, with the companies concerned. And this, of course, has consequences in respect of both oil pricing and tax calculations — as follows:

(i) Oil pricing is a key element in determining tax liability. This is because the price declared (received) for the oil produced determines the value on which PRT calculations are based. It also helps to determine corporation tax liability. However, the nature of the world oil market is such that unless oil is sold at 'arm's length' its value is not an objectively determinable element.[1] The market for most internationally traded oil is essentially one of negotiated transfer prices (between one company and another of the same group and between companies and government in respect of most OPEC, and some non-OPEC, oil-producing countries) in which the value of a barrel of oil to one company is different from its value to a second or a third company. Similarly, the value attached to a barrel of oil is different as between companies and governments — depending on factors of overall supply/demand situations within corporate and other entities and in the light of the infrastructural contrasts and the different market openings etc, of the various parties. In this sort of situation, almost all governments depending on, or expecting, revenues from oil production have found it necessary to abandon regulations which required, in effect, that each barrel of oil be separately valued, by agreement between the company concerned and the government or its fiscal agency. In order to safeguard revenues such governments found that it was essential to define, unilaterally, a set of tax reference prices as the sole means of valuing the oil produced. Such tax reference pricing is required in respect of the UK's oil production in order to replace the valuation

1. For a detailed discussion of the point see 'The Oil Companies and the New World Oil Market' in the *1978 Yearbook of World Affairs*, University of London Institute of World Affairs, London, 1978.

procedure as set out in the Oil Taxation Act of 1975. Although this Act allows the Inland Revenue to substitute its assessment of market value for that declared by the licensee in respect of sales between affiliates, this can be challenged on appeal by the company concerned. And this creates a situation which must serve to inhibit the valuation assessment by the Inland Revenue.[1]

(ii) The contrasting value of certain quantities and qualities of oil to different parties at specific times and specific places creates a possible 'conflict' between the best interests of the government and the commercial interests of BNOC. The latter, in order to enhance its potential profitability, must seek as low a price as possible from the companies for its purchases of participation oil, as well as for oil it purchases from elsewhere, in a situation in which it is acting as a crude oil trader with oil bought from other parts of the world. Such 'low' price levels could then be used by the oil companies in support of claims for low valuations of their North Sea production. This could be to the detriment of the government's interests in terms of maximising its tax take. The institution of a regulated tax reference price system for offshore oil production would, of course, also serve to eliminate this potential conflict between the government and the state oil corporation. It would, in addition, eliminate the latter's potential embarrassment in possibly having to modify its efforts to secure a 'bargain' because of the effect this could have on government interests.

1. Note that a difference of $1 per barrel between a higher tax reference price and a lower so-called market (transfer) price will, at end-1979 rates of North Sea production, make up to £375 million per year difference to the UK's balance of trade figures and up to £300 million difference to government revenues (given that the higher cash flow through higher revenues will comprise mainly additionally collectable taxes). A difference of $1 per barrel is by no means high: in a dispute between the US government and some major oil companies over the costs of moving oil via the trans-Alaska pipeline, the difference between the parties was over $4 per barrel. I have heard the argument, but have not been persuaded by it, that a tax reference price for British oil would upset our tax agreement with the US. The US Treasury has accepted tax reference prices for oil in other countries for many years and I see no inherent reason why the UK should be exceptional. However, given the large and increasing amounts of export earnings/revenues involved, even if there is an exceptional situation, the present method whereby the value of oil is determined for tax purposes seems worthy of examination for changes which could be advantageous to the UK.

(iii) The evaluation and the incidence of costs is, of course, also critical for the tax calculations — as recognised in the Oil Taxation Act of 1975 and its Schedules. Yet we do not, and indeed cannot, know if it is being monitored/controlled effectively. The question of allowable costs, in respect of specific operations for a specific oil company, is a confidential one for determination by the Inland Revenue alone. This is done on the basis of the general rubric whereby secrecy for individuals' tax returns is maintained. Is it reasonable, in view of the very large revenues expected by the government from a limited number of companies specifically allocated rights to exploit part of the nation's oil resources, and in the context of a highly complex technico-economic situation with many international aspects, that the tax assessment should be secret? Given the nature of the operations and the very few companies involved, ought not the process of tax assessment be open? In his 1977 Energy Plan, President Carter insisted that there must be openness in the evaluation of the oil companies' responsibility for paying taxes. He added that this must be done in respect of each individual operation by a company and that every such operation (ie each producing field, each refinery etc) should be treated **separately for tax purposes. The United States has, relatively, a much lower degree of financial dependence on the** oil tax-take and a much higher degree of domicility in that country of the oil companies concerned. It also has a much longer experience of trying to tax oil companies effectively. In view of this background, the fact that it considers that oil companies must be subject to special rules over tax assessments emphasises the need for such a requirement in the case of the UK.[1] This general proposition may be supplemented by two specific needs in respect of monitoring/control over the development of

1. The importance of the requirement is emphasised by the current high degree of uncertainty concerning future oil revenues. Although the widely varying views on future revenues (for example, estimates for the year 1983-84 range from £4750 million to £8250 million) relate, in part, to uncertainty over levels of both prices and production, they are also a function of contrasting expectations on the degree to which the company will actually have to pay PRT and Corporation Tax. The higher estimates fail to recognise the opportunities which the present system gives for the tax minimisation procedures.

continental shelf oil under the existing pattern of concession/development legislation, *viz:*

First, for each oilfield development plan, the companies are required to indicate capital and operational costs and their expectations of inflation in respect of both. How, and to what degree, are the companies' expectations in these respects related to the expenses they ultimately claim year by year against their Petroleum Revenue Tax (PRT) obligations on individual fields?

Second, PRT was made chargeable on a field by field basis to prevent the obligation 'disappearing' into operations elsewhere in the offshore area. This has not been done in respect of Corporation Tax which may be payable by a company in respect of its earnings on a particular field. Thus, a company with a near-future obligation to pay Corporation Tax can avoid doing so by making an investment elsewhere in the UK offshore region at the appropriate time. Whilst this is in keeping with the normal rules for calculating any company's liability for Corporation Tax and whilst, in respect of offshore oil operations, it may be a 'useful' device in order to get new fields developed so as to provide continuity in oil flow later in the century, is it reasonable that this should be achieved at the expense of taxes from offshore operations that would otherwise be payable in the short term? This not only discriminates against other existing taxpayers in general, it also discriminates, in particular, against those oil companies which did not make an early strike of developable proportions. And this in itself was not a matter of luck or competence — it emerged out of the preferences shown to certain companies in the early discretionary awards of blocks in the North Sea!

(d) The inherent inflexibility in the application of the concession system, as it has been developed, and the accompanying highly complex associated system of taxation, seem to have produced an unwillingness and/or an inability on the part of the companies which have made 'small' finds in the North Sea to develop them. But the designation of a small field as one with less than 100 million barrels or even one with less than 200 million barrels is absurd. Elsewhere in the world of oil, such fields are considered as 'moderate' or even of 'large' size and are eagerly developed. Compare the reaction of the companies to fields of this size in the UK offshore to the recent reaction by Exxon to its discovery of a very modest field (by North Sea standards), estimated to contain some 50 million barrels, in the US Gulf of Mexico. The field, moreover, has been discovered in no less than 1,200 feet of water on the continental slope so it can hardly be said to be in a friendly environment. Indeed this depth of water compares with 600 feet or less in the North Sea. This offsets, in part at least, the latter region's harsher meteorological and water conditions. Yet, in spite of this and in spite of the fact that the cost to Exxon of developing the field concerned is front-end loaded by the $71 million which it had to pay for the small concession (small, that is, by North Sea block standards), the company moved with speed towards its development. Dozens — perhaps even hundreds — of discoveries of this size have been, or can be expected to be, found on the UK Continental Shelf. If they are not developed, then the country's oil potential will be seriously diminished and there will be consequential severe economic disadvantages.

If the nature of the concession and its associated taxation system, as it has evolved to date for Britain's offshore resources, eliminates the possibilities of the development of this oil, as seems to be the case (in spite of the tax concessions which have been or which could be extended to small fields under existing legislation), then the validity of the system, as an appropriate tool for the continental shelf's exploitation, must be seriously in doubt.

(e) In other words, there remain major uncertainties over the degree and the speed of the future exploration and exploitation effort on the UK's Continental Shelf. The way in which the system works also makes it impossible to forecast the overall impact of offshore oil and gas developments on the British economy and society in general, and on specific geographical areas in particular. This is largely because of the incalculable returns from the development in respect, specifically, of government revenues from the taxation system. Continuing modifications to the system, emerging out of the realisation that it is not producing the benefits which were expected, seem likely to fail to make any *essential* difference to the situation and the outlook.[1] My recommendation, therefore, is that this conventional, concession-style approach should be abandoned as the basis for organising the exploration and exploitation of the rest of the UK Continental Shelf's oil and gas resources: in the same way as it has been abandoned in every other major oil-producing country outside North America (where special conditions, especially the prior payment of bonuses for concessions, apply). It should be substituted by the establishment of a new type of politico-economic environment for the exploration and exploitation of the country's remaining offshore oil resources. In case this should not be thought to be worthwhile because so much has already been done in the North Sea, it should be noted that most of the UK's Continental Shelf and adjacent areas of the continental slope have yet to be explored. (See frontispiece map.) Thus, even if the remaining exploration is only modestly successful, the future development of British oil will take place over a period several times longer than the decade and a half of development we have had to date. It is therefore not too late to consider alternative systems which seem inherently more likely to ensure both full exploration of the resources and appropriate benefits to the country from the continued development of offshore oil and gas.

1. In addition to the points made above, there is an additional one which can now be made in light of the decision in 1979 to increase the rate of PRT from 45 to 60 per cent and to reduce from 175 per cent to 135 per cent the 'value' of capital investment in a field's development to set against PRT calculations. We have re-run those parts of our computer programmes relating to the development of the Forties, Piper and Montrose fields (see Odell and Rosing, *The Optimal Developments of North Sea Oilfields*, Kogan Page, London, 1976) in which we measure the NPV of government tax-take in the systems as developed, and find that the 'improvements' in government revenues arising from these apparently very significant changes in the tax regime is less than 12 per cent over the life of the fields. And this 'increase' is dependent on the assumption that all Corporation Tax, due from each field's development, will be paid and not postponed, perhaps indefinitely, as a result of the investment of all or part of the profits from a specific field's development elsewhere in the North Sea. Thus, it seems that even formidable changes in the tax rules under the existing system do not make striking differences to the expectations. This is true even in respect of fields already under development and of investment in which, therefore, the companies concerned have committed themselves as a result of a decision based on the previous, lower rates of tax. With the higher tax rates it is likely that a less intensive development plan for a particular field would have been chosen in order that the company could maintain its required rate of return on investment. This would have served to reduce the level of government revenues on the fields concerned; a danger in respect of new developments given the recent 1980 budgetary decision to increase PRT yet again, this time to 70 per cent.

D. An alternative environment for the exploitation of the rest of the UK's offshore oil and gas resources

1. The auction system

(a)   Though this is also a system based on concessions it is, nevertheless, one which is essentially different from the discretionary system. The market mechanism (expressed through the opportunity which potential exploration companies are given to evaluate the worth of a particular area) is the regulator of developments and the means whereby the state attempts to collect — in advance — the economic rent (= the supernormal profit) which the companies bidding for a concession expect to emerge from the exploitation of the reserves.

The arguments in favour of the auction system have been presented generally in a recent publication by Professor K W Dam[1] and its application to the UK Continental Shelf has recently been recommended by Professor C Robinson and Dr J Morgan[2] and in a paper published by the Bow Group of the Conservative Party.[3] It is thus conceivable that the auction system could be chosen as an alternative basis for future UK policy towards oil exploration and development.

(b)   I am not persuaded of its economic and politico-economic validity in general and, in particular, I am not convinced that it is appropriate in the case of the UK Continental Shelf if the government's wish to maximise its return from offshore oil and

1. See K W Dam, *Oil Resources: Who gets what how?*, University of Chicago Press, Chicago, 1976.
2. In C Robinson and J Morgan, *North Sea Oil in the Future*, MacMillan for the Trade Policy Research Centre, London, 1978.
3. P Lilley, *North Sea Giveaway: The Case for Auctioning North Sea Oil Licences, A Bow Paper*, Bow Publications, London, 1980.

gas developments is to be realised. There are two conditions which are necessary to make it possible for the auction system to secure all or most of the economic rent obtainable from oil and gas exploitation, *viz:*

(i) that there is effective competition between many applicants for concessions in the auctioning process;

(ii) that there will be no significant changes in the economic environment subsequent to the allocatory decisions made through the auction.

(c) I have presented my conclusions on these — and related — issues elsewhere.[1] In brief, however, my view is that these necessary conditions are not, and indeed cannot, be met in respect of the further development of the UK Continental Shelf.

(i) Even in the United States it is only with some difficulty that the oil industry has been kept on the straight and narrow path of competition for exploration rights. This has involved complex Federal and State legislation and a plethora of regulations which have been evolved over a long period of time. However, outside the US there is a general lack of experience in dealing with the mainly American oil companies and the latter are, as a consequence, in a much stronger position to assert their control or influence over any 'objective', essentially laissez-faire allocatory procedures. Given that the completion of these allocatory procedures then leaves the state bereft of influence over what happens next, in respect of the exploitation of any oil discovered, the auction system seems inappropriate for the UK, where efforts over the last ten years have been devoted mainly to evolving a post-discovery set of regulatory conditions. All this experience would be wasted and a new start in regulating the allocatory process would have to be made if an auction system were now to be adopted.

(ii) In the US the oil and gas industry has been generally iso-

1. These can be found in a review article I wrote on Professor Dam's book (op cit) in *Energy Policy*, Vol 5, No 3, September 1977, pp 256-7.

lated from the conditions which apply in the rest of the world — or, rather, in the world of the international oil companies. In particular, companies bidding for concessions in the US can be very confident, not only that they will have a market for any oil found, but also a market at what is virtually a guaranteed price. Thus companies with their prior knowledge of these important economic variables can more easily determine what it is reasonable to bid for a concession. Elsewhere in the world, however, including the UK, oil markets have been and remain open to competition from international oil, so that companies are only able to bid a low figure for a concession — in order to minimise their risk.[1] Indeed, outside the US the auction system seems unlikely to be able to secure for any one country a set of auction bids which is even as high as the world-wide average level of economic rent expected from all new developments by the companies.

In the case of the UK, therefore, the only reasonable auction bids which could have been made by the companies, before 1974, would not have secured any part of the much higher economic rent which has emerged out of the increases in the price of oil since then. In such circumstances an auction system would soon have produced bad company/government relationships.[2] This remains the case, given that the world-wide uncertainty over the supply and price of oil continues. Thus, the apparently simple and self-regulating 'auction system' cannot be expected to work effectively for an equitable and effective development of the rest of the oil resources on the UK Continental Shelf.

1. Note that the bids received for the few British blocks which were auctioned in one of the earlier rounds were derisory compared with bids made for acreage on the US Continental Shelf by the same companies.

2. This has even been the case recently in the US where 'old oil' (ie oil from old concessions) has become so profitable for the companies, in spite of lower oil prices in the US, that the government has had to introduce a special petroleum windfall profits tax — and, in so doing, has produced the worst oil company/government relations in the industrialised world!

2. A discretionary system of concession allocations — with high royalties

(a)   The lower than expected revenues to the government from the development of the North Sea oilfields, which are already in production, appear to arise from the ability of the companies to adjust their policies over prices and over the allocation of costs, so as to minimise their liabilities to PRT and Corporation Tax. In view of this and if it were thought desirable to maintain the essential framework of the concessionary system as it has been developed, then there is much to be said for calculating the approximate royalty rate which is equivalent to the 70 per cent rate of PRT, and for charging that rate of royalty on production (additional, of course, to the 12½ per cent royalty already levied). This will provide the government with a much more easily calculable — and an inherently less avoidable — level of tax income. The ease of the calculations and the non-avoidability of the amount calculated by the government as due from the companies, would be enhanced if this approach were also combined with a tax reference price valuation for all oil produced.

(b)   A royalty rate of, say, 50 per cent (inclusive of the existing 12½ per cent royalty) on the value of the crude produced is one which would give the government 75 per cent of the total profits earned from a field, if it is assumed that costs account for about 33 per cent of the cash flow. What the application of a single rate of royalty would not do, of course, is to distinguish between fields and parts of fields in terms of the variable costs of their development. The result would be that some companies would restrict their developments to the lowest cost fields and to the lowest cost development systems on larger fields, and so still make 'too much' profit. Meanwhile other companies would risk earning too few profits if they tried to maximise the production of oil by including investment in production facilities with a low productivity. The system would also fail to allow for the way in which costs, relative to prices, change over time and so penalise those companies with a high cost-to-price relationship, whilst favouring those in the reverse position.

(c)   The Province of Alberta in Canada is one major producing area in which a high royalty rate has been introduced as the means whereby the government collects the bulk of its income from oil exploration and development. The new royalties are basically much higher than they were prior to 1974 (up to 44.33 per cent compared with a maximum of 25 per cent previously). A system has now been introduced whereby the actual rate varies from month to month and from well to well and also according to whether the oil is 'old' or 'new'. The new complex scale starts as low as 8.866 per cent and can, as indicated above, be up to five times as great.[1]

(d)   The Alberta formula, involving a well-by-well calculation, could not be directly applied to an offshore area like the UK Continental Shelf where the producing platform and its associated wells, rather than individual wells, is the basic operational unit. It does, however, show what is achievable through a royalty system. Its relevance to a UK situation is certainly worthy of more detailed examination, particularly in respect of the areas which have already been allocated and where it would be impossible to change fundamentally the conditions of development.

(e)   It must be noted, however, that there are several reservations arising from this approach to state involvement in oil exploitation. The state certainly achieves efficiency in collecting its revenues, but the system (except in respect of the initial decision to offer concessions) does not imply much else by way of government involvement in the development. In particular, in the case of the UK Continental Shelf developments, it would raise the question of the status of BNOC in that a high royalty system is not really compatible with direct state involvement in the exploration and development process. If the state entity had to pay the much higher royalties required then it could be inhibited from undertaking nationally desirable activities, such as the 'non-commercial' search for additional reserves. If it didn't pay them, then the charge of unfair competition from the private companies would clearly be justified. It also leaves the question of a private company's liability to corporation tax unresolved[2] and one must recognise that, except when the

1. See Michael Crommelin, 'Government Management of Oil and Gas in Alberta', *Alberta Law Review*, Vol XIII, No 1, 1975, pp 146-211.

2. In the case of Alberta there is no corporation tax as this is not a Canadian provincial tax right. There is, however, a Federal Tax on oil companies' profits (as on company profits in general). There seems to be no inherent reason why the royalty and the corporation tax should not be levied in a complementary way by the same authority.

royalty is taken in kind, the system still requires attention in respect of the oil valuation question.[1]

3. The inadequacies of the concession system — even with a state oil company

(a) The nub of the criticism of the concession system and its derivatives lies in the style of relationship between companies and government which it necessarily implies. It is basically a system in which all the positive initiative — except the granting of the concessions in the first place — lies with the companies, to whose actions the government can respond only in a neutral or a negative way. It is, moreover, a system in which the state's collection of revenues from the exploiters of the resources is seen as a 'burden' on the company. In other words, it is a system which, in politico-economic terms, hardly seems compatible with the concept of the national ownership of a country's oil and gas resources.

(b) The relationship does not, moreover, change over time. A growing number of regulations certainly serve partially to block loopholes through which potential revenues leak away. However, it also creates an environment in which the concessionary companies continually react against the more stringent conditions by finding new ways of avoiding the worst of their consequences. Thus, in the final analysis, the government does not make the progress it expected to make in controlling the system. Nor does it benefit, as calculated, from sharing in the profits of oil production. It is, in essence, a system of thwarted expectations on the part of the government — and of equally thwarted efforts on the part of the companies which often cannot make the progress they expected, because of the suspicions to which they become subjected by the very workings of the system.

(c) This necessarily unsatisfactory outcome of the concession approach to the development of oil and gas resources has already been partially recognised in the UK. The solution

1. The price of oil month by month is one of the variables written into the formula for calculating the monthly basic royalty rates on Albertan oil.

sought has been that of the creation of the British National Oil Corporation which, under the recent Labour Administration, had to be involved in all new concessions with a minimum equity interest of 51 per cent. This, of course, gave it a controlling influence over exploration and development, with the intention of providing a counter-balance to the problem of the lack of positive direct government involvement in the concessions. It provides a solution to the problem, however, only insofar as one assumes that the interests of the state company and of the state remain identical. Given that BNOC is required to act 'commercially' when in partnership with private oil companies, this assumption does not seem to be justified. If not, then the mere creation of a state company to work alongside the private companies does not in itself provide the whole answer to the problem posed for the state by the concession system.[1]

(d) Moreover, in one very important way the existence of a single state entity like BNOC, with an interest in all the concessions, acts as a barrier to the most effective development possible in a large, varied and geologically complex set of opportunities for oil and gas exploration and exploitation as offered by the UK's large offshore areas of continental shelf and slope. This does not have anything to do with the level of technical competence of the state entity or with the arguments on state *v* private enterprise approaches to oil development. It arises in a much more simple way, *viz* from the fact that a single interpretation must prevail in any organisation about the significance of geological, geophysical and/or drilling information for a region's oil and gas development potential. However, given the existence of many schools of thought on the chances for oil and gas occurrence in respect of all the many different kinds of habitats for hydrocarbons, there is a danger that many good opportunities will be missed because of the limited range of views which any single organisation (no matter how big) can embrace. Indeed, given the necessary hierarchical structure of any exploration and production division in an oil company (state or private) there is likely to be a well-defined, accepted

1. This does not imply, of course, that there are no ways in which a state oil company modifies the state's relationship with the oil companies in a concessionary system. Of particular importance is the 'information' aspect, whereby more effective regulations can be achieved.

'company view' of what is, or is not, worth pursuing. As a result, many quite reasonable opportunities will be ignored or dismissed by any single entity and the development of the total oil resource base will be less than complete.[1] Thus an organisational structure for oil and gas exploration whereby this danger can be avoided, through a 'plurality' of exploration and development efforts, is a 'must' for a system within which there is direct state involvement in the exploration and development system. This suggests that the stage now achieved in the institutional approach to the development of oil in the UK's offshore areas ought not to be viewed as other than an interim one. The full and profitable development of oil and gas resources depends not only on the replacement of the concession system, but also on a new form of state/oil companies' co-operation which is better able to ensure that the geological complexities of the vast offshore areas in the North Sea and elsewhere on the UK Continental Shelf and slope which remain to be explored, will indeed be tackled. (See frontispiece map.)

4. An alternative institutional arrangement: a production sharing system involving joint venture operations.

(a) Any part of the continental shelf and slope (involving either one block or contiguous blocks) should be open for a bid for an exploration/drilling licence at any time by any party which sees it as offering an opportunity for potential development.[2] This in no way diminishes the responsibility of

1. There is evidence that this has already happened with BNOC. Following the views of its chief geological adviser, BNOC appears to 'believe' that all the large North Sea fields have been found and that two-thirds or more of the region's oil resources have been discovered. This attitude must, of course, be a dominant influence in BNOC's exploration/development policy and, as BNOC has been required to be the majority shareholder in all recent concessions, it must also be a dominant influence in the future exploration history of the province. Yet its view of the potential for further North Sea development is by no means generally accepted. Other companies think there are many large fields and extensive resources still to be discovered (eg Philips). With a single state company view as the only one possible, alternative ideas on the size and nature of the resource base risk never being put to the test — to the country's disadvantage, not simply in respect of its oil sector, but also in more general politico-economic terms.

2. The number of bids accepted for consideration and determination at any point could, however, still reflect both policy and practical

the Department of Energy as the final arbiter of what is appropriate in respect of the speed and intensity of development, but
it does provide an environment in which the many possibilities
for successful exploration are not foregone. Whether or not the
party making the bid seems to be 'capable' of undertaking the
task[1] ought not to be a prime consideration in the government
response to the bid. An incapable entity will not be able to
pursue its intentions very far — and we ought to be prepared to
allow such a company to 'lose' its own shareholders' money in
respect of the work it asks to be allowed to do in the country's
offshore areas. Even a company which proves to be incapable
of sustaining an exploration and/or a development programme
will generate some employment and income in the economy,
produce some geological knowledge and, possibly, generate an
element of technological advance in exploration methodologies.
Thus there will still be a benefit to the community from such
private enterprise failures.

(b) The bids, however, should *not* be for a concession to
explore for and, if successful, to produce oil, on the value of
which the company which made the bid will eventually pay
taxes (as well as relatively minor sums in respect of lease rentals
etc). Instead, a company which is successful will then bid for
the right to retain a share of the production of the oil which it
has discovered and hopes to produce. The balance of the oil
to be produced (and which will normally be the greater amount,
usually by a very significant margin) will remain the property of
the state (or a designated state company) for it to do with as it
pleases in a basically 'no-pay' situation. The company, in other
words, bids for the privilege of spending its own money on
exploring an offshore prospect (with one or more specified
exploration wells). If it is not successful with the wells as designated, then the agreement either lapses immediately or possibly
may be extended if both company and the Department of
Energy agree to more drilling. If the drilling is successful, then
development of the field — as initiated and/or approved by the

considerations. The latter involves the ability of the Department
of Energy to handle and process applications; the former implies
concern for the overall rate of exploration/exploitation in the
light of national needs, in relation to exploration successes to
date and the likelihood of 'required' rates of production being
achieved.

1. Except, of course, in terms of a technical ability to undertake
drilling in a safe way.

Department of Energy — will take place[1] within the framework of a joint venture between the company concerned and a state entity. The latter will be responsible only for a previously agreed part of the costs. This is a matter for *ad hoc* negotiations by the three parties concerned (company, state entity and Department of Energy) and it *may* be a nil percentage. The state entity will later be responsible for handling and selling that part of the production which is retained by the state — from a minimum of, say, ± 50 per cent (in respect of difficult-to-develop and/or small fields) to 80+ per cent for the lowest-cost-to-develop and/or better located fields.

(c)   This partnership system thus entails a series of government/oil companies' joint ventures based on the concept of production-sharing. Each agreement involves the negotiation of the specific joint venture arrangements and implies agreement on the division of the production of oil, as well as on the nature of the developments to be undertaken, and on the division of responsibility for the costs. The company bidding for the venture will, of course, pay all the exploration costs and it may offer to meet most or even all of the development expenditure. Thereafter, government regulation of the venture at the politico-economic level is minimised — it will relate simply to questions involving the development plan for the field or fields[2] but there will be no problems of valuing the oil for royalty calculations and/or for special petroleum taxes. Similarly, as the latter will not exist, there need be relatively little government concern for evaluating the costs which are incurred in developing the field. All or most of the costs will be purely for the account of the company, as agreed in the negotiations.

(d)   The state's essential economic interest lies in the share of the oil and gas it retains from the development, and the disposal of which becomes the responsibility of the appropriate state entity — say, BNOC in respect of oil and BGC in respect of

1. Should the drilling be successful, but not 'successful enough' in the company's view to justify development, then all interests in the block will automatically revert immediately to the Department of Energy which may, of course, then receive other bids for the acreage or it may decide to require development by a state entity.
2. Government regulations of a technical character in respect of oilfield practices and environmental considerations etc will, of course, still be required.

natural gas. The extent (if any) to which a state entity partici-
pates directly in a specific development can also be a matter for
negotiation and there is, of course, no reason why the existing
public hydrocarbon corporations cannot take the initiative in
deciding to seek the right from the Department of Energy to
explore/develop a particular prospect. In such flexible circumstances there would seem to be no reason whatsoever for any
fixed upper or lower limits to the degree of state involvement —
though general government policy requirements will influence
the overall degree of state participation which is sought from
time to time. Nevertheless, even with a minimum element of
direct state involvement in exploration and production there
could still be a significant presence by the national oil and gas
corporations in this sector of the oil industry. The system
would certainly provide the opportunity for the state corporations to develop their expertise and knowledge of the industry.
If it is thought desirable, they could broaden their activities,
both functionally and geographically, as is indicated below.

(e)   This production-sharing, joint venture approach to the
development of Britain's offshore oil wealth does not lie in the
state having a specific share — say, 51 per cent — in each development in order to institute an element of control into an otherwise eminently uncontrollable concession style system of the
traditional pattern. Under the joint venture, production-sharing
agreement the state automatically and directly retains ownership
and control over most of the oil produced. This directly provides
the equivalent of the royalties and the special taxes of the concession system without the state having to be concerned for the
intermediary values of production costs and the 'value' of the oil.

(f)   The state achieves an immediate and a continuing access
to a minimum of about 50 per cent and up to a maximum of
perhaps 85 per cent of the oil produced. How the government
— as opposed to the state entities — will assume its interest in
financial terms is a matter for study. Perhaps the state entities
might be permitted to retain enough of the value of the oil[1] to
cover their costs (plus a margin as a 'handling commission')
with the balance going to a National Oil Account or its equivalent. The role of the state entities would now be much

1. The valuation for this purpose would be by the state. There would
   be no interests of the private companies to take into account.

more orientated to the functions of handling the oil — with up to perhaps 70 per cent of the total amount of oil produced at any time as the initial responsibility of these organisations.[1] In such circumstances the state entities will be major sellers of crude oil on the international market, or they could sell to the oil companies, or become significantly involved in downstream activities — with consequences and opportunities which are discussed below.

(g) Under the joint venture, production-sharing system as described, the private companies secure access to the crude oil available as a return on the privileges extended to them for finding and producing oil from Britain's offshore areas. Most of the companies involved are multi-national oil companies with international supply, transport and refining operations, the overall optimisation of which the companies seek to achieve through complex computer-based methods, without any particular concern for the specific national interests of the countries in which their operations are located. It therefore may be considered appropriate for this joint venture type of approach to be extended to the transport, refining and associated activities of the oil industry in the UK.

(h) In the first place the joint venture concept could be extended to the transportation of the crude oil produced through common-user pipelines. Each of these lines could thus be jointly developed by several companies, including one or more of the state companies, which would be responsible for handling most of the oil from any group of fields as a result of the production-sharing agreements on the fields concerned. Such common-user joint venture lines will eventually form part of an offshore pipeline system, the need for which emerges out of the increasing number of fields under production. The development of these ought primarily to be related to the system's overall economics, rather than to the needs of a particular field. The arguments in favour of a unified approach

1. The use of the plural here — and elsewhere — in the discussion of state entities reflects the immediate possibility of both BNOC and the BGC being involved in the new system. It also indicates the future possibility of new state oil entities being created as a mechanism for ensuring a preferred multi-faceted public interpretation of the geological opportunities on the continental shelf (*see* footnote on p 36).

to the offshore transportation system, in order to avoid the unnecessary duplication of facilities and/or the isolation of some fields which are capable of being developed from an economic means of transport, have already been presented above (*see* pages 20 and 21). They are mentioned again at this point in order to show how a common-user pipeline system fits more easily into a joint venture, production-sharing system. As the government, through one or other of its entities, automatically has the responsibility for at least 50 per cent of all crude oil produced — and usually a much higher percentage — then a common-user, offshore pipeline system is an almost inherent part of the production-sharing system recommended in this report.

(i)   Under existing legislation for the exploitation of Britain's offshore oil resources, the state has assumed few powers over the oil once it has been landed in the UK. There is, indeed, nothing more than the 'expectation' that up to two-thirds of all offshore oil production shall be refined in the UK. The Pipelines and Submarine Pipelines Act of 1975 was concerned in part with refineries (Part IV) but only in the context of controlling and authorising their construction and extension by private companies. Authorisation has to be 'consistent with the national policy relating to petroleum' — but 'national policy' was not defined in the Act. The same Act (Part I) also extended to BNOC the powers 'to provide and operate . . . refineries in connection with petroleum' but such power depends on the specific consent of the Secretary of State (*see* page 21). The Act, moreover, gives no indication that the power to provide refineries was to be sought as a means of changing the pattern of oil refining and distribution which emerged in the UK between 1945 and 1974 under the stimulus of a rapidly increasing demand for oil products in the energy and chemical sectors of the economy. A development which was, of course, related to the use of imported oil.

(j)   However, the optimal development of the country's offshore oil resources appears to necessitate a fundamentally re-structured oil refining and distribution system. The essential elements in this re-structuring are presented in the next part of this report. Meanwhile, one can summarise as follows what appear to be the advantages for the UK from the introduction of a production-sharing system, in place of the concession system, in respect of the areas of the continental shelf and slope which remain to be explored and exploited for their oil and gas resources:

(i) It ensures the necessary and continuing availability to the UK of a broad range of oil companies' exploratory and development expertise whereby the probability of the successful discovery and production of the remaining offshore oil and gas resources can be maximised.

(ii) It provides the opportunity for definitive, firm and continuing government control over the speed and location of developments — within the context of the strategy determined to be most appropriate for any given period.

(iii) It ensures a 'clean' and a guaranteed flow of benefits to the government from the development of oil- and gasfields, thus greatly reducing the uncertainty prevalent now over the size of the flow of benefits. This is a result of the present complex system of concessions, regulations and taxation in which the important cost and value variables, so influential in determining the government tax-take, cannot be readily or easily forecast. The companies, under the production-sharing agreements, also know exactly where they stand in respect of each agreement which they sign.

(iv) It gives the government, through its state entities, access to most of the crude oil produced.

(v) In that most of the oil produced is not transferred to private ownership as in the exploration/exploitation system developed to date, there is a firmer basis for a rational pattern of offshore transportation and for the further evolution of a rationalised refining/distribution system.

(vi) With ownership of most of the produced oil remaining directly in the hands of the state, there is also a firmer base for Britain's use of oil as a 'weapon' in negotiations over energy and over other policies within the EEC; and with other nations or groups of nations.

(vii) The inherent flexibility of the production-sharing system (each joint venture agreement is negotiated individually in the light of the specific circumstances of the discovery and the specific interests of the parties concerned at a particular moment) means that an effective interest in small fields and in the economically less attractive parts of larger fields, can be generated through the negotiations between the state and the company or companies concerned. The system implies a joint state/company evaluation of the prospects for each field discovered and hence a compromise acceptable to both parties is more likely to be reached than in the present much more rigid system. At present the commercial evaluation of the field is completely divorced from the government's evaluation of its interest.

(viii) It establishes a framework for a significant new style of oil company/government relationship which is also applicable in many other parts of the world — notably in the still largely unexplored potentially petroliferous regions of the Third World. The development of these areas is of great importance in future overall oil supply/demand relationships and the experiences gained, with the framework as defined, could provide the UK with the opportunity to play a leading role in encouraging and in making possible the development of those resources.

E. The integration of the downstream activites of the UK oil industry into the production sharing system

1. The oil industry in the UK — before North Sea developments

(a) The British oil industry, until a few years ago, consisted of refineries, product import terminals, inland pipelines and road/rail/canal terminals, distribution facilities, and, of course, gasoline stations and other wholesale and retail outlets. It would be inappropriate to describe the system in this report and I have, in an earlier publication, attempted to describe, analyse and define possible policy options in respect of the government's role in that system.[1]

(b) Before North Sea oil was discovered and even with the low oil prices of the pre-1973 situation, there were elements in the UK's oil industry which seemed to require state intervention. It was needed to ensure the rational use of resources, fair prices and the elimination of unnecessary developments in the industry's infrastructure which is, of course, notoriously disadvantageous for the environment. At that time I suggested that there was a need for a 'National Oil Agency' with particular responsibilities. These would have included 'the control and direction of British oil import policy; the control and direction of refining and distribution within the country; ordering the systematic exploitation of whatever oil wealth may be found in UK offshore areas; and representing Britain's interest in negotiations with other nations of the world with an interest in oil'.[2]

(c) The role I suggested for that Agency in respect of the

1. See Peter Odell, *Oil: The New Commanding Height*, the Fabian Society, Research Series Pamphlet No 251, December 1965.
2. ibid, p 24.

country's offshore oil wealth are now vested in the Department of Energy and/or BNOC and this report has been mainly concerned with indicating ways and means whereby the 'systematic exploitation' of this wealth can be more effectively achieved. The likelihood of continuing success in that direction, however, now makes it even more important than it was previously to pay attention to 'oil import policy' and to 'refinery distribution'. These aspects of the oil industry's activities are not unrelated to Britain's role as a major new producer of the commodity. Indeed, I would put the requirement for new import and refining policies etc in a much more positive way, *viz* that part of the benefits to the country, arising from its role as an oil producer, are dependent upon changed attitudes and policies towards oil imports, and to refining and distribution. It would be stretching the main subject matter of this report were I to give *detailed* attention to these downstream sectors of the now much expanded UK oil industry. There are, nevertheless, some major issues which it is essential to define and appraise in order to complete an analysis of the opportunities and challenges to government arising from the development of oil production in Britain's offshore waters.

2. Oil imports, refining and distribution — with Britain as a major oil producer

(a)   The oil sector of the UK economy has traditionally been free from government intervention. This was increasingly inappropriate as oil inexorably became the commanding height of the energy economy in the years between 1951 and 1971. Today, the absence of effective government interest in, and concern for, the industry's downstream activities could lead to restraints on the oil production sector and create the possibility of wasted investment.

(b)   There are several structural reasons for this — organisational and geographical, as well as political and economic. They derive from the fundamental changes in the country's oil supply position. These changes create new and unconventional problems arising from the ownership and location of the oil industry's importing, refining, transportation and distribution facilities in relation to the availability of oil from the North Sea, sufficient not only to meet domestic demand but also to provide

some oil for export. In essence, the shares of individual companies in the UK oil market — and in its refineries and other
infrastructural elements — do not match (and are not likely to
match for the foreseeable future) the company by company
availability of UK-produced oil. For example, refineries which
could run in large part on North Sea oil cannot do so because
they are owned by companies with inadequate North Sea production. On the other hand, companies with a large supply of
crude oil from North Sea fields, which they have developed, do
not have access to domestic processing facilities and domestic
markets. Even where the problems are not as clear-cut as this,
there are, nevertheless, difficulties which arise from the con-
trast between the historic locations of companies' refining and
other facilities, and the requirements for handling and processing
crude oil supplies originating in the extreme north-east of the
country.

(c) These infrastructural problems, in respect of both the
organisational and geographical aspects, could lead simul-
taneously, in the short to medium term, to high imports of
expensive oil by companies lacking North Sea supplies on the
one hand and, on the other hand, to restraints on production
from their North Sea fields for companies lacking sufficient UK
crude oil outlets. Indeed, by the mid-1980s the former companies may be forced by the OPEC nations, from which they
draw the bulk of their supplies, to take specified packages of
crude oil and oil products whilst the latter group of companies
is forced to shut in potential North Sea oil production. These
potential illogical developments in the UK oil system could,
however, be sorted out within the context of a joint venture,
private companies/state entities' approach to the organisation
of the downstream operations of the oil industry in the United
Kingdom.

Given such an approach, the use of UK refineries for UK-
produced oil could be divorced from the question of the particular company by company ownership of the facilities. In
addition, an overall rationalisation of the supply, refining
and distribution system could be sought. Within such a framework, the relationship between oil exports and oil imports —
both crude oil and products — could be established so as to
maximise the return to the nation from, for example, the
appropriate blending of North Sea crude oil with inherently

lower-value imported crude. It would also provide a realistic way of ensuring that the 'expected' up to two-thirds of UK-produced oil is, indeed, refined in the UK. Assuming, of course, that this 'expectation' emerges as the most beneficial from an industry-wide evaluation of the opportunities available within the context of the recommended state/private enterprise, joint venture approach to downstream operations. In this context, the refineries of the UK could be utilised (with appropriate investment in both expanded and refurbished facilities) to their optimum level. This would produce favourable results from the point of view of the unit cost of refining, whilst the most economic crude oil-to-refineries and refineries-to-distribution centres transport facilities could be developed to the advantage of the oil consumer. These advantages would be expressed in terms of lower refining and transportation cost elements in the price paid for the oil products needed and thus also benefit the oil companies involved.

(d)   This rationalisation process should be a joint state enterprise/private enterprise venture. Judging from the way oil companies have to evolve, it is clear that BNOC cannot, for very much longer, avoid becoming involved in some way in the oil industry's downstream operations. But, given the proposed integrated state enterprise/private enterprise joint venture system, BNOC ownership of refinery capacity need not *necessarily* be affected. Instead a re-structured BNOC could, if desired, simply become a co-ordinating authority for the most appropriate pattern of refining operations. Such co-ordination, however, seems likely to be more effective if the state entity were to have a direct involvement in refining and other downstream activities. This could be achieved in one of several ways, *viz:*

(i)   By BNOC being taken into equity and operational partnership by one or more of the existing companies with refining interests.

(ii)   By BNOC buying an existing refinery from a company willing to sell.

(iii)   By BNOC taking over one or more refineries from a company or companies that did not wish to be involved in a joint venture approach to this sector of the industry.

48

Its direct ownership of refinery facilities, moreover, need not necessarily prevent BNOC from becoming a participant in all the refineries in a way parallel to its participation agreements with the successful concessionary companies in their North Sea oilfields.

(e) The justification for this sort of development arises from the fact that the UK, like all other oil producing and exporting countries, will inevitably have to make its refining (and even its petrochemical) industries an integral part of its oil-producing sector. In the not too distant future, in respect of the longer-established oil producing and exporting countries, this seems certain to mean that crude oil exports will only be allowable within the context of an overall oil export policy which requires a greater or lesser emphasis (as determined from time to time) on the exports of oil products! The UK, as an oil exporter, will have to follow suit in order to protect its interests from being undermined by other oil exporting countries. Thus, any country seeking to import crude oil from the UK would also be expected to take a specified (though variable) percentage of its overall import needs from the UK as oil products.

(f) The location of the oil refining industry was historically and traditionally linked with the production of oil. However, because of factors specific to the international oil situation in the 1950s and the 1960s this locational link between oil production and refining was broken, so that many refineries were established (as were the UK's refineries) in centres of oil demand. The reversion of the general refinery location pattern to the *status quo ante* is now, in my view, only a matter of time. If this is so, then there is no inherent reason why the excess refinery capacity of Western Europe should be adjusted downwards to overall current demand levels on an equal, or a nearly equal basis, for all the countries in the region. Within the European Economic Community, given that the supply of a not inconsiderable part of its oil needs is now likely to be met by oil produced from the continent's own offshore resources, then those member countries, such as the UK, with an availability of indigenous crude oil, must be expected to increase their share of the total refinery throughput that is required by the overall level of oil demand in the EEC as a whole. In this context, the United Kingdom has a readily justifiable case for an EEC refinery policy which recognises the UK's new loca-

49

tional advantages for refinery development.

(g) Thus, as far as the UK is concerned, the expansion of its downstream oil activities is part of the justifiable use of its oil as an element in its economic policy-making at the EEC level. There are, needless to say, limitations on the use of oil in this way. If they are pushed too far, then the partner countries may look elsewhere for their oil import needs and so limit the UK's opportunities in respect of its oil wealth. It is this negotiable situation which provides the basis for a reasonable and realistic oil policy by the UK *vis a vis* the other member countries of the EEC. The development of Britain's considerable offshore oil resources provides the basis for economically rational linkages into expanding downstream activities. The use of these will also benefit the Community, in terms of a guaranteed supply of an essential commodity at prices which could also be negotiated to ensure advantages both for the UK and the rest of the EEC. In this respect, too, a positive policy, by the UK government, as far as oil refining, transport and distribution are concerned, would serve to enable the UK to take full advantage of its offshore oil. It would also serve as a means whereby the problem of surplus downstream capacity in the EEC as a whole, is divorced from the complications of company supply patterns and put into an economically rational framework for solution by the governments concerned.